AI AWARENESS SERIES

AI in Agriculture
and Food Production

Andrea V.M. Greaves

Contents

Introduction

Welcome to AI in agriculture and food production, part of our AI awareness training series. This short book is designed to give you a clear, practical overview of how artificial intelligence is transforming the way we grow, manage and deliver food across farms and greenhouses.

AI helps monitor crops, improve yields and even automate planting and harvesting. From precision agriculture to autonomous machinery, smart systems are making agriculture more efficient and responsive. We'll explore how AI assists in crop monitoring, supporting farmers with real time insights into plant health and quality. You'll also see how AI is used in livestock management and food processing, helping to streamline operations, maintain hygiene and ensure quality.

Further along the food chain, AI supports supply chain, logistics, optimizing how produce is sorted, packed and delivered. We'll look at sustainability too, how AI tools contribute to regenerative agriculture and climate aware farming practices. And throughout you'll see how AI powered platforms give farmers better decision making, tools helping them adapt to changing conditions and global challenges. Let's get started.

Chapter 1: AI in Crop Monitoring

Modern crop monitoring plays a vital role in improving yields and ensuring food security. By tracking crop development closely, farmers can detect issues early—such as pests, diseases, or nutrient deficiencies—before they escalate. This leads to more targeted use of resources like water and fertilizer. And, importantly, it supports sustainable farming by minimizing waste and reducing environmental impact.

Artificial Intelligence is reshaping agriculture by providing highly precise monitoring of both crops and the environment. It automates complex data analysis, turning raw inputs into fast, actionable insights. This not only reduces manual labor but also boosts accuracy, allowing farmers to make better decisions more efficiently.

When AI is integrated with traditional crop management methods, the result is a powerful hybrid approach. AI enhances the precision and efficiency of farming without discarding local expertise. This balance supports both innovation and the sustainability of cultural agricultural practices.

Satellite imagery allows us to monitor large-scale agricultural areas with consistent coverage. These images help detect crop stress early—often before it's visible to the naked eye—allowing for quicker interventions and better overall crop health tracking over time.

Drones add a layer of precision by capturing high-resolution images of specific areas within a field. This helps farmers identify localized issues and take targeted action. With drone data, decisions around irrigation, pest control, and fertilizer use can be made with much greater accuracy.

AI processes the massive volumes of image data collected from satellites and drones. This enables quick extraction of useful insights, like identifying weak areas in the field or detecting signs of pest activity. AI can also assess overall crop vigor, making it easier to tailor interventions for maximum yield.

Machine learning models are trained to recognize early signs of crop diseases using both image data and environmental variables. Catching symptoms early allows farmers to respond before the problem spreads, potentially saving a harvest.

AI-powered systems can also automatically detect and classify pests. This means farmers can apply pest control measures more precisely—treating only affected areas and reducing chemical use. The AI can differentiate between pest species, making control strategies even more targeted and effective.

One of the biggest advantages of AI in pest and disease detection is speed. Quick identification of problems leads to lower treatment costs, reduced crop loss, and improved productivity. This responsiveness is key to maintaining healthy crops and sustaining farm income.

Chapter 1: AI in Crop Monitoring

AI algorithms analyze crop growth data along with external variables like weather and soil conditions to predict yield potential. This gives farmers valuable foresight, helping them monitor development in real time and respond proactively to environmental stressors.

By integrating past crop data and environmental conditions, AI can make more accurate yield forecasts. This includes looking at long-term weather patterns and past yields, giving farmers a data-driven way to plan for the future.

Accurate yield predictions help farmers use resources more efficiently—from fertilizers to labor. They also improve supply chain planning by enabling better timing for harvesting and distribution. This minimizes waste, reduces spoilage, and supports a more efficient farm-to-market pipeline.

Chapter 2: Automated Farming Systems

AI integration in agricultural machinery is revolutionizing farming. AI-powered navigation allows machines to move precisely across fields, improving efficiency and reducing the need for manual labor. Obstacle detection systems help avoid collisions and ensure safe operation. And machine learning models continuously optimize tasks, adapting to environmental changes over time.

Automated planting and harvesting bring several major benefits. They reduce labor costs by handling tasks that would otherwise require a workforce. They also increase both the precision and speed of operations, making sure planting and harvesting happen at the best possible time. And by optimizing equipment use, they help improve overall crop yields and minimize losses.

Let's look at some real-world examples. Farms around the world are using AI-powered equipment to boost efficiency and reduce reliance on manual labor. These tools have led to noticeable increases in productivity and better resource management. Plus, automation has helped cut operational costs by streamlining processes with AI-driven systems.

Robotic weed detection relies on advanced sensors that identify weeds early on, enabling timely action. Computer vision algorithms analyze images to distinguish weeds from crops with high accuracy. Once identified, robots either mechanically remove the weeds or treat them, reducing the need for herbicides and making farming practices safer for the environment.

Pest monitoring is also being automated. Using cameras and sensors, these systems track pest populations in real time. When treatment is necessary, pesticides are applied in a targeted manner, minimizing chemical use. This targeted approach also helps prevent pests from developing resistance over time, making pest control more effective in the long run.

Robotic weed and pest control technologies bring clear environmental and cost benefits. They reduce the amount of chemicals needed, promoting sustainability. Automation also lowers labor requirements, making farming less intensive. And by minimizing both crop loss and treatment costs, these systems improve overall cost efficiency for farmers.

Inside autonomous greenhouses, sensors play a key role in monitoring environmental factors like temperature, humidity, CO_2, and light. Automated systems use this data to adjust ventilation, heating, and lighting automatically, creating ideal growing conditions without constant human oversight.

Greenhouse automation doesn't stop at climate control. Robots handle seed planting with high precision, ensuring uniform growth. Automated systems keep a constant check on plant health, catching problems early. And when it's time to harvest, robotic systems work efficiently, improving consistency and productivity while reducing the need for manual labor.

Smart irrigation systems start with continuous soil moisture monitoring, ensuring crops get just the right amount of water. Weather sensors also track conditions like temperature and rainfall, providing critical data. Together, these systems enable irrigation schedules that adjust dynamically in response to changing environmental factors.

Automation makes precision watering possible. Automated irrigation systems deliver water directly where it's needed most, using methods like drip and sprinkler systems. This precise control minimizes waste, ensuring that every drop of water benefits the crops effectively.

The impact of smart irrigation technology is significant. By optimizing water use, it helps conserve water resources — an increasingly important goal for sustainable farming. These technologies make it possible to address water scarcity challenges while promoting responsible and efficient farming practices.

Chapter 3: Livestock Management

Traditional livestock management has heavily relied on manual observation. Farmers and handlers would watch animals closely to assess their health and behavior. Decisions on feeding and breeding often depended on experience and intuition rather than data. However, these traditional methods face limitations—they don't easily scale to larger operations and may not always be accurate.

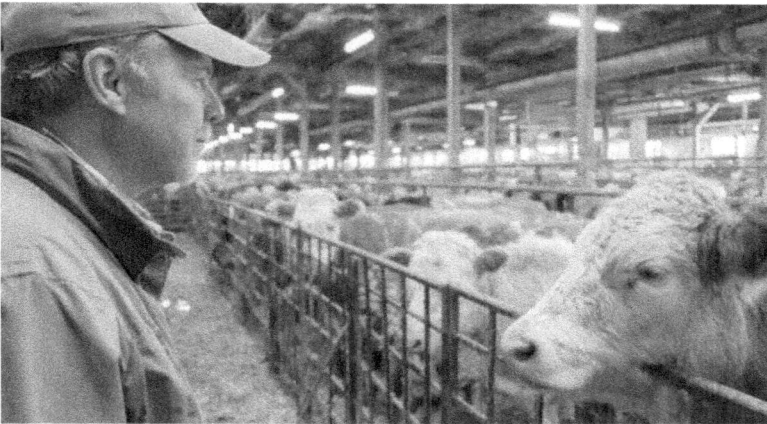

Technology-driven solutions are changing livestock management. With sensors, farmers can now monitor livestock health and environmental conditions in real time. Artificial intelligence takes this a step further by analyzing collected data, providing valuable insights for better decision-making. Automation also plays a big role—helping to streamline tasks, improve efficiency, and reduce manual effort on the farm.

Modern livestock management brings several benefits. Animal welfare improves because of better monitoring and proactive care. Technology boosts productivity and helps reduce costs. However, there are challenges too—like high upfront investment costs, data privacy concerns, and the need for specialized technical skills to manage advanced systems.

Let's start with AI for health monitoring using sensors. Wearable health devices can track vital signs in real time, helping detect illness or distress early. Temperature monitors and movement sensors like accelerometers help assess well-being by tracking body heat and activity levels. GPS trackers monitor livestock location, which is particularly useful for managing grazing and ensuring safety.

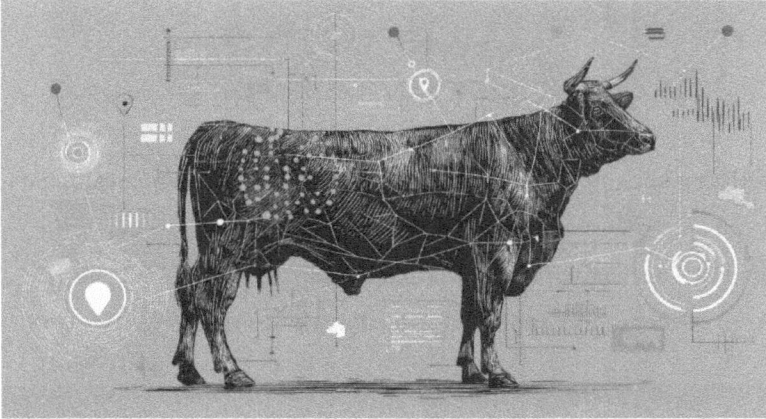

Sensors collect real-time health data, giving farmers continuous updates on their animals' conditions. AI processes this data immediately, spotting patterns and detecting any deviations from normal health indicators. This real-time analysis enables proactive health management, allowing farmers to respond quickly to potential problems.

AI plays a key role in early disease detection. By analyzing health data, AI can spot the early signs of disease—often before they become obvious. This enables quicker diagnosis and treatment, reducing the spread of illness and improving overall herd health and welfare.

Facial recognition technology uses advanced image processing to capture and analyze the unique features of each animal's face. AI algorithms compare facial data for reliable identification. This method is non-invasive, meaning animals don't need physical tags or invasive procedures to be recognized.

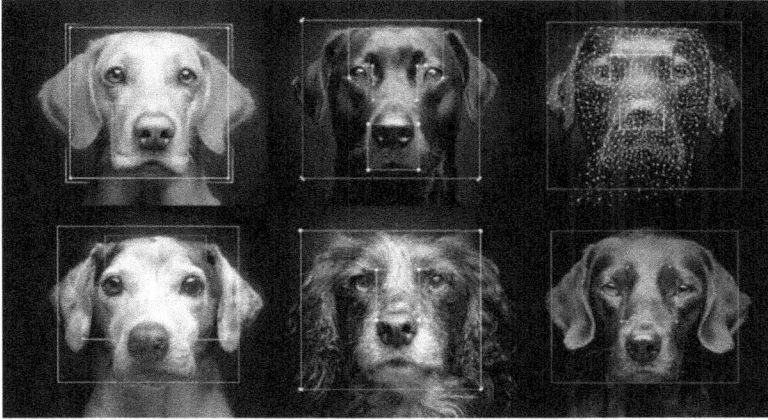

Facial recognition enables automated identification of animals, which is extremely useful for farm management. It also helps maintain accurate health records, improving care and disease control. Additionally, tracking breeding and movement history becomes easier, supporting better management decisions and enhancing productivity.

Accurate identification leads to better record-keeping, which is crucial for managing livestock effectively. It also supports compliance with legal regulations and standards in livestock management and trade. Furthermore, strong identification systems improve supply chain traceability, adding transparency and ensuring safety from farm to market.

AI techniques also help monitor animal behavior. Machine learning algorithms analyze animal movement to detect both normal and unusual patterns. AI also monitors feeding behavior, which can reveal health issues. Even social interactions among animals are tracked, giving insights into group dynamics and potential behavioral problems.

AI systems can detect anomalies in animal behavior—like reduced activity or irregular feeding—which may signal health problems. Detecting these issues early allows for prompt intervention, improving animal welfare and reducing risks to the herd.

Behavior analysis through AI allows for the early detection of health or stress-related issues. Farmers can apply targeted preventive measures based on these insights, helping reduce risks and promote better overall livestock well-being.

AI plays a significant role in feed optimization and weight prediction. Data-driven strategies help create precise feed formulations. AI models can predict weight gain and monitor growth, providing valuable insights to guide interventions and optimize animal growth.

With AI predicting weight and growth patterns, farmers can make timely adjustments to feeding and management practices. This leads to healthier animals and better growth outcomes, enhancing productivity and profitability.

AI-driven feed optimization improves efficiency by preventing overfeeding and reducing waste. This doesn't just lower costs—it also helps minimize the environmental impact of farming. Plus, proper feed management supports animal health by ensuring balanced nutrition.

Chapter 4: Climate-Aware Farming

Artificial intelligence is transforming weather forecasting for agriculture. By analyzing large, complex weather datasets, AI models can detect patterns that traditional methods might miss. This leads to more accurate forecasts with longer lead times, giving farmers critical information to anticipate weather conditions and make better planning decisions for their operations.

Precise weather forecasts offer several key benefits for farm operations. They allow farmers to schedule planting, irrigation, and harvesting at the best possible times, improving yields. Accurate forecasts also help reduce losses from extreme weather events, protecting crops and assets. Ultimately, this improves both the efficiency and productivity of farming through informed, timely decision-making.

When AI-driven weather data is integrated into farm management systems, it enhances decision-making with real-time environmental information. Automated alerts can notify farmers immediately about weather changes, helping them adjust operations quickly to protect crops and resources. With these systems, farmers can make real-time adjustments to irrigation, planting, and harvesting activities for optimal outcomes.

Identifying local climate risks is a crucial first step for farmers. Different regions face varying threats such as droughts, floods, and extreme temperatures, all of which can impact crop yields. Conducting thorough climate risk assessments helps farmers plan proactively, enabling them to manage operations more effectively and reduce vulnerability to climate events.

Developing adaptive strategies is key to managing extreme weather risks. Diversifying crops can spread the risk and improve resilience. Adjusting planting dates can help avoid adverse weather conditions. Improved water management, like efficient irrigation, supports both drought resilience and flood protection. And by adopting resilient farming practices, farmers can maintain productivity even in challenging climates.

Real-time decision-support tools offer farmers timely insights during extreme weather events and other critical situations. These systems help farmers make informed decisions on the spot, protecting crop yields and conserving vital resources. The ability to react quickly with accurate information is essential for modern, climate-aware farming.

Data analytics plays a powerful role in guiding crop choices. By analyzing historical and projected climate data, farmers can select crops that are well-suited to local conditions. Considering soil quality and market demand also strengthens decision-making. This data-driven approach boosts the chances of higher yields and improved profitability.

Farmers also need to adapt to changing climate patterns. Selecting crop varieties that match evolving temperature and rainfall trends can enhance resilience. Adjusting planting schedules to align with new climate realities helps reduce risks and improve crop success rates.

Let's look at some case studies. Data-driven crop selection has been shown to increase resilience to climate variability. Farms that have optimized their crop choices see higher yields and better resource use. This approach also supports sustainable farming practices by lowering environmental impact and conserving key resources.

Measuring agricultural carbon emissions is essential for sustainable farming. This includes tracking fuel use in machinery, monitoring

fertilizer application to account for nitrogen emissions, measuring soil carbon changes, and assessing greenhouse gas emissions like methane and nitrous oxide. Together, these measurements give a complete picture of a farm's carbon footprint.

There are several sustainable practices that can help reduce agricultural emissions. No-till farming reduces soil disturbance and improves soil health. Cover cropping protects the soil, boosts nutrient retention, and enhances carbon sequestration. Using fertilizers precisely minimizes runoff and emissions. Finally, integrating renewable energy sources like solar and wind cuts fossil fuel use and lowers overall emissions.

Accurate emissions reporting and sustainability certification can make a big difference. Digital platforms allow farmers to report their emissions transparently, supporting better climate-friendly decisions. Certification helps build trust with buyers and opens up access to green markets that prioritize sustainable products. These tools give farmers both environmental and economic benefits.

To conclude, advancing sustainable agriculture depends on effectively using data and technology. By adopting climate-aware practices—like AI forecasting, adaptive strategies, data-driven crop selection, and carbon tracking—farmers can improve resilience, enhance productivity, and support a healthier planet.

Chapter 5: Farm Data Platforms and Decision Support

Agricultural data comes from many sources—sensors, satellites, weather stations, and farm equipment. These sources track multiple parameters critical to farming operations. By integrating all this data into centralized data lakes, we make it consistent and accessible for analysis. This centralization directly supports better decision-making and leads to improved productivity on the farm.

Knowledge graphs allow us to connect different agricultural entities— like crops, soils, weather patterns, and farming methods—into a structured network. By building these relationships, farmers gain richer insights into how these factors interact and impact yields. Knowledge graphs also enhance querying capabilities, allowing users to discover hidden patterns and correlations that traditional databases might miss.

By harmonizing data from diverse sources, we ensure that different farming systems can communicate seamlessly. This interoperability boosts efficiency and coordination. With unified data, farmers make better-informed decisions, optimizing yields and resource usage. Ultimately, this approach supports both productivity and sustainable farming practices.

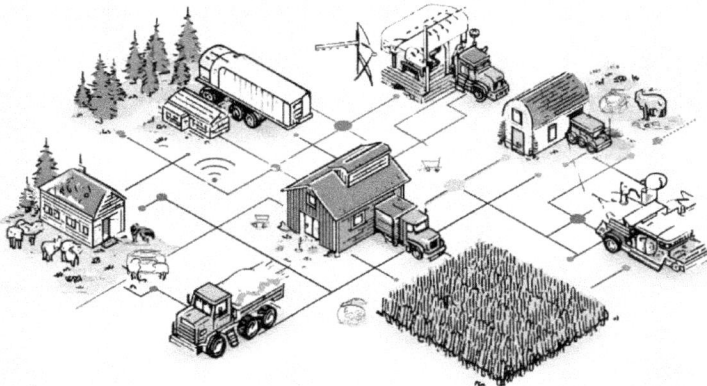

Predicting crop yields is crucial for planning. Advanced models use environmental data and historical trends to forecast yields accurately. Similarly, market trend forecasts help farmers align their production with market demand, maximizing profitability and reducing waste.

Analyzing weather patterns helps farmers anticipate risks—like droughts or storms—and schedule operations more effectively. Assessing soil health guides planting and fertilization decisions. Monitoring environmental factors further supports risk management and helps farmers identify opportunities to boost crop success.

Scenario modeling lets farmers explore "what if" situations before committing to a strategy. It helps them assess how different conditions—whether environmental or economic—might impact their results. This tool is especially valuable for making sound decisions even when future conditions are uncertain.

Prescriptive AI takes decision-making a step further by optimizing schedules for irrigation and fertilization. AI-driven precision ensures that crops get exactly what they need, when they need it. This reduces resource waste, supports environmental sustainability, and enhances crop health by maintaining optimal growing conditions.

Beyond managing water and nutrients, AI also helps assign labor and equipment dynamically based on real-time needs. This not only cuts costs but also ensures that farming operations are more efficient and environmentally sustainable.

Efficient resource use directly translates to lower operational costs and better environmental outcomes. By minimizing waste—whether water, chemicals, or energy—farmers protect ecosystems while maintaining productivity. Prescriptive AI helps achieve that critical balance between profitability and sustainability.

Mobile devices like smartphones and tablets enable farmers to collect data directly in the field, improving accuracy and timeliness. This real-time data capture enhances situational awareness, allowing farmers to respond quickly and make better-informed decisions on the spot.

With Edge AI, data can be analyzed on-site, without relying on a central server or constant internet connection. This means farmers can get instant diagnostics—like pest detection or nutrient issues—and receive actionable recommendations right in the field, enabling prompt intervention.

Real-time insights allow farmers to monitor and respond to field conditions as they happen. They can adapt their operations—whether planting, irrigating, or harvesting—on the fly, minimizing risks from sudden weather changes or pest outbreaks. This rapid adaptation leads to better crop performance and improved resilience.

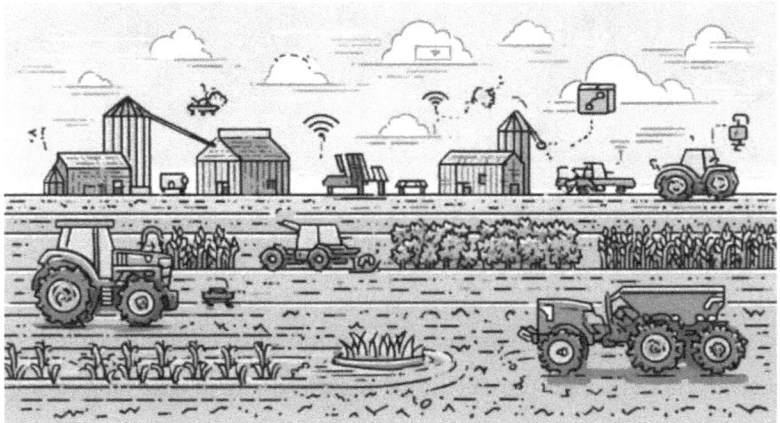

s.

Chapter 6: AI in Agricultural Supply Chain Optimization

AI plays a vital role in agriculture, transforming how farmers manage crops and operations. With crop monitoring, AI can detect plant health issues early, helping to prevent losses. For yield prediction, AI analyzes environmental data to forecast crop yields, enabling farmers to make informed decisions. Automated machinery powered by AI ensures efficient planting, watering, and harvesting. And in supply chain management, AI optimizes logistics and distribution, reducing waste and improving overall efficiency.

Why is supply chain optimization so critical in agriculture? First, it helps reduce operational costs, making farming more profitable. By streamlining processes, we also minimize food waste from production to distribution. Ensuring timely product delivery means fresher produce reaches consumers, which is key in agriculture. Finally, a well-optimized supply chain can boost farmer incomes by improving market access and reducing losses.

Adopting AI in agricultural supply chains brings clear benefits—like improved forecasting and logistics optimization, which enhance efficiency and reduce waste. However, there are challenges. Poor data quality, complex system integrations, and a lack of skilled personnel can all hinder AI adoption. It's crucial to address these issues head-on to fully unlock AI's potential in agriculture.

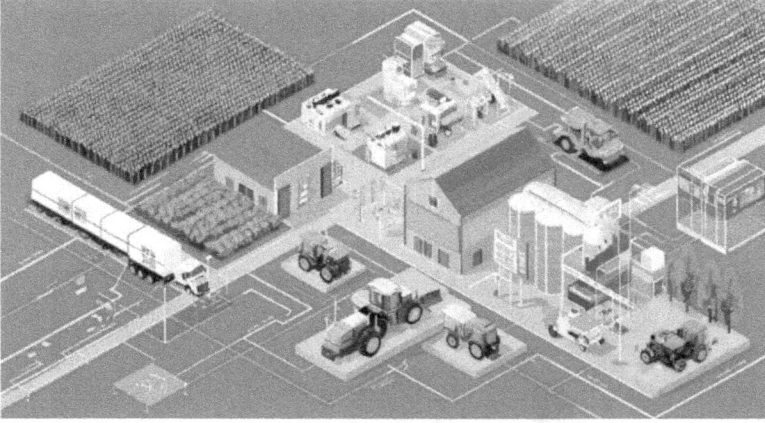

AI improves demand and supply forecasting by analyzing both historical and real-time data. By studying past crop yields, weather patterns, and market trends, AI identifies patterns that influence forecasts. When combined with real-time updates—like current weather conditions and market inputs—AI can dynamically adjust predictions, making them more accurate and useful for planning.

Artificial intelligence also helps predict market trends and consumer behavior. It analyzes purchasing patterns to detect demand shifts and evaluates social trends to anticipate changes in consumer preferences. With these insights, businesses can adjust production schedules and distribution plans to meet market needs more effectively.

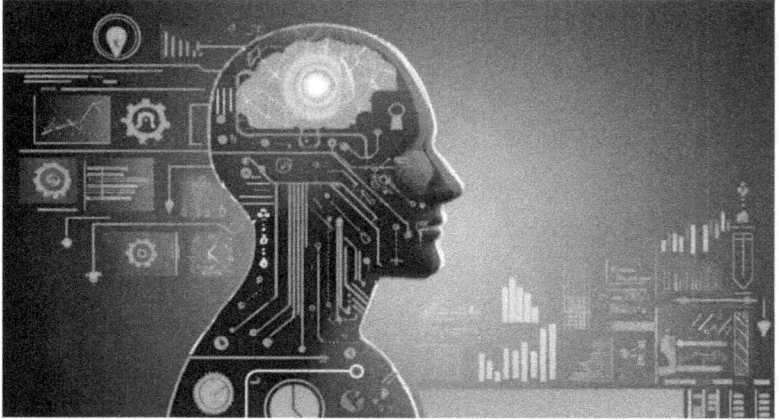

Accurate demand forecasting is crucial for managing inventory and reducing waste. AI enables businesses to maintain optimal inventory levels—preventing overstocking and reducing spoilage. This leads to less waste throughout the supply chain, supporting sustainability goals and delivering cost savings.

AI-driven models significantly enhance route planning in agricultural logistics. By analyzing real-time traffic data, AI helps avoid congestion and improve delivery times. Algorithms also calculate the shortest possible routes to minimize travel distance and reduce fuel consumption. Additionally, route planning can take delivery windows into account, ensuring timely arrivals and higher customer satisfaction.

Maintaining product quality is especially important in agriculture, and AI-based cold chain monitoring plays a key role. AI continuously monitors temperature and storage conditions during transport, helping prevent spoilage and maintaining freshness. This ensures that products arrive at their destination in optimal condition.

There are several success stories showing how AI improves logistics and reduces costs in agriculture. AI optimizes delivery routes, cutting down transportation time and expenses. Cold chain monitoring systems powered by AI help maintain the right conditions for perishable goods. Together, these applications not only save money but also enhance the overall quality of agricultural products.

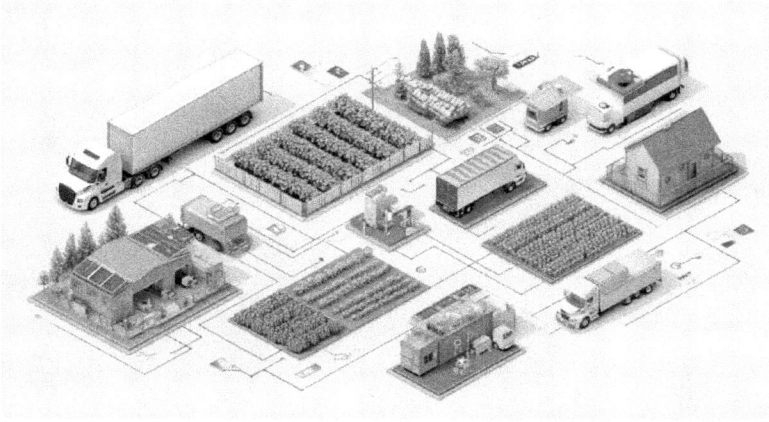

AI analytics can detect critical points of loss within the supply chain. By monitoring data, AI identifies where spoilage or damage occurs most frequently. This information helps businesses target interventions at those points, improving efficiency and reducing waste throughout the supply chain.

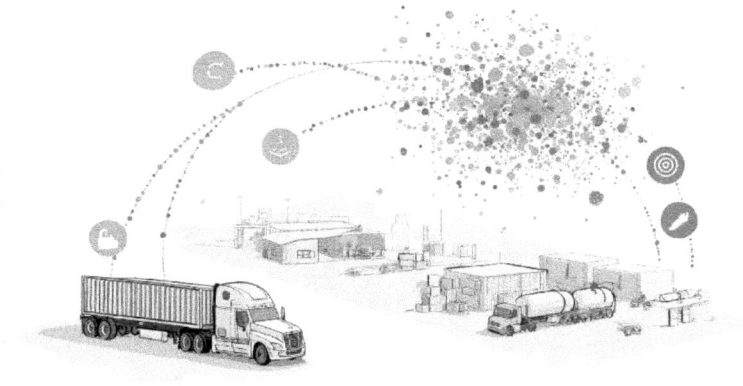

Predictive maintenance is another area where AI adds value. By anticipating equipment failures before they happen, AI helps maintain reliability and avoid costly downtime. Automated quality control systems also play a role by detecting product defects early, reducing waste and improving the quality of goods moving through the supply chain.

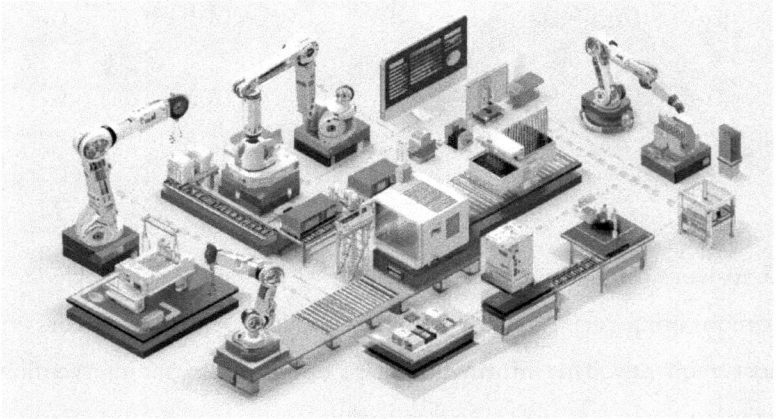

AI can enhance storage and transportation strategies in agriculture. It monitors storage conditions and adjusts them as needed to preserve product quality and extend shelf life. For transport, AI plans routes and schedules that minimize delays—ensuring products are delivered on time and in the best possible condition.

The integration of IoT devices and AI enables complete traceability across the supply chain. IoT devices collect real-time data at every stage, from farm to retail. AI then processes this data, providing transparent tracking and ensuring accountability throughout the supply chain.

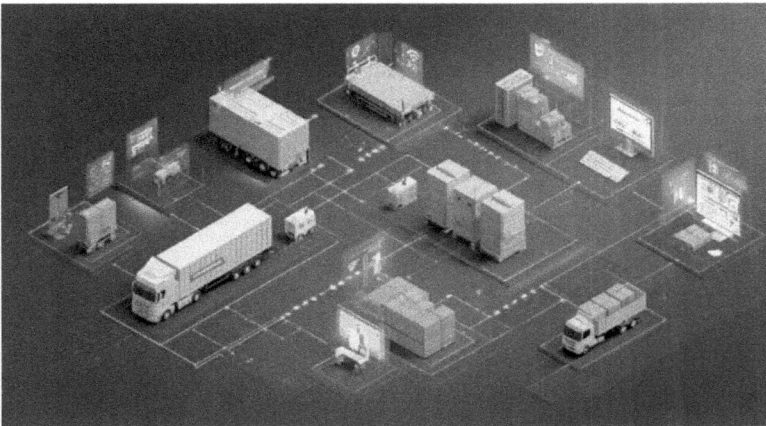

Real-time data analytics offers significant benefits for all stakeholders. With up-to-date insights, businesses can respond quickly to changing demand. They can also identify and mitigate risks before they escalate, making operations more efficient, reducing costs, and boosting productivity.

Real-time visibility into the supply chain empowers businesses to make quick, informed decisions. With immediate insights, they can detect disruptions and act swiftly to resolve them. This capability strengthens supply chain resilience, allowing companies to adapt better to market changes and unforeseen challenges.

Chapter 7: AI for Food Safety and Quality Control

Artificial intelligence is reshaping the food industry in several key ways. Machine learning helps with predictive analytics and automation, making food production both safer and more efficient. Computer vision allows automated inspection, spotting defects in real time. And data analytics helps organizations ensure compliance with food safety regulations by processing large amounts of data quickly and accurately.

Ensuring food safety and quality comes with challenges. First, raw materials vary a lot, making it tough to guarantee consistency. There's also the constant risk of contamination, so monitoring needs to be strict. Rapid inspection is critical — we need to spot problems fast. And traditional methods often fall short because they rely heavily on manual checks, which are slow and prone to mistakes.

AI-driven solutions offer major benefits. They enhance accuracy, improve efficiency, and help detect risks early. With AI, the food industry can move towards more reliable and proactive quality control systems that protect both businesses and consumers.

AI vision systems start by capturing high-quality images of food products. The AI then analyzes features like size, shape, color, and texture. This analysis helps grade and sort items consistently, making sure only quality products move forward in the production process.

In real-world settings, AI vision systems sort fruits, vegetables, and processed foods by detecting defects and classifying items into quality grades. These systems can automatically remove substandard products, which improves overall quality and safety. And because it's automated, the process is much faster and requires less manual labor.

Compared to traditional inspection, AI vision offers several advantages. It's more accurate, reducing errors and defects. It works faster, increasing production speed. It's non-destructive, so products aren't damaged during inspection. And by removing human bias and fatigue from the process, AI ensures reliable, consistent results every time.

Predictive modeling plays a key role in food safety. By analyzing historical and current data, AI models can estimate the risk of microbial contamination. This allows for proactive risk assessments and supports early intervention, reducing the chances of foodborne outbreaks.

Several types of AI models are used for microbial risk assessment. Neural networks are great at spotting complex patterns. Decision trees break down decisions into simple steps. Support vector machines help categorize risks precisely. And ensemble methods combine different models for even better accuracy.

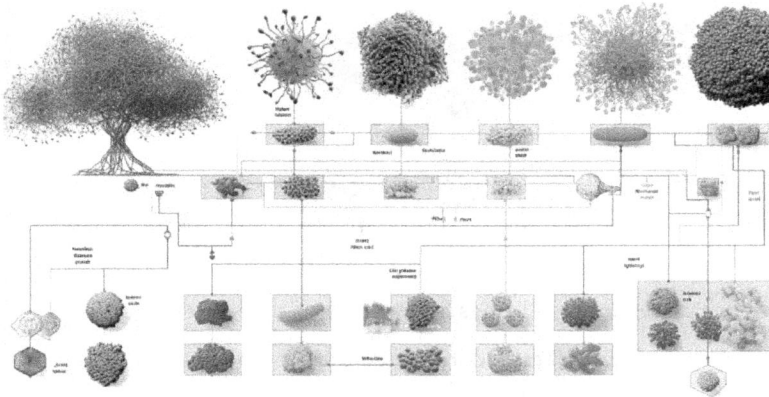

AI-powered risk prediction is already making an impact in the food industry. Companies use AI to forecast microbial risks, leading to fewer contamination incidents. These models also help businesses meet strict food safety regulations more easily, ensuring both compliance and consumer safety.

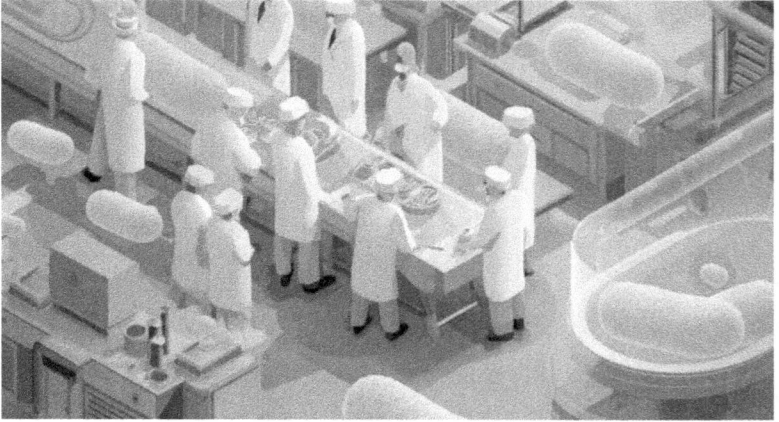

Blockchain technology brings a new level of transparency to food supply chains. Every transaction is recorded permanently and can't be changed, ensuring data integrity. This allows every stage of the production and distribution process to be tracked, making it easier to trace the history of a product.

When combined with AI, blockchain becomes even more powerful. AI can analyze blockchain data to detect unusual patterns and predict potential risks. This integration allows for automated decisions that improve efficiency and accuracy while giving companies better tools to monitor their supply chains in real time.

Traceability has big benefits. It allows fast identification of contamination sources, which helps prevent health risks. It also boosts consumer trust by making product information transparent. And when recalls are needed, they can be managed more efficiently, protecting both public health and company reputation.

Sensors and data analytics are critical for spoilage detection. Advanced sensors can monitor gases, temperature, and humidity — all key indicators of spoilage. Data analytics then processes these sensor readings, spotting spoilage patterns before they become a problem.

AI algorithms analyze sensor data to recognize patterns and detect anomalies. By catching these anomalies early, companies can take action before spoilage becomes visible, protecting product quality and reducing waste.

By using AI to monitor spoilage, companies can extend product shelf life and enhance food safety. Early detection means fewer health risks and less food waste. AI also helps optimize inventory and storage management, making supply chains more efficient and sustainable.

Chapter 8: Sustainable and Regenerative Agriculture

Let's start by defining sustainable and regenerative agriculture.

Sustainable agriculture focuses on meeting today's food needs without compromising the environment or the ability of future generations to do the same.

Regenerative agriculture goes a step further—it's about actively restoring soil health, boosting biodiversity, and supporting the ecosystem as a whole.

Sustainable and regenerative agriculture is essential for environmental health.

By maintaining soil fertility, we help preserve the nutrients and structure that crops need to thrive long-term.

Reducing pollution means using fewer harmful chemicals and managing farm waste better, which protects both land and water ecosystems.

Conserving water through efficient techniques supports farming and preserves natural water sources.

And by enhancing biodiversity, we create more resilient ecosystems that can better withstand environmental pressures.

However, there are significant challenges in adopting these practices.

Farmers often face knowledge gaps—new technologies and ecological insights can be complex and not always easily accessible.

Financial constraints are another hurdle, as sustainable methods and advanced tech can require upfront investment.

And finally, managing complex ecological systems takes expertise, making it challenging for farmers without specialized support.

Chapter 8: Sustainable and Regenerative Agriculture

Let's talk about crop rotation and cover cropping—two key sustainable farming strategies.

Crop rotation involves alternating the types of crops grown, which helps maintain soil nutrients and naturally reduces pests.

Cover cropping means planting crops in the off-season to protect soil from erosion and add organic matter, which improves soil health.

AI-driven decision support systems are changing the game for farmers.

These tools analyze data and offer tailored recommendations on crop rotations and cover cropping strategies, helping farmers make smarter, more sustainable decisions.

AI offers real benefits for both soil health and yield.

By optimizing moisture retention and aeration, AI helps improve soil structure.

It also boosts nutrient availability, ensuring plants get the minerals they need.

AI can reduce pest populations with more targeted interventions.

And overall, this leads to higher crop productivity and more efficient food production.

AI-powered remote sensing tools, like drones and satellites, are vital for modern farming.

These tools collect high-resolution environmental data from the air and soil.

AI processes this data quickly, giving farmers insights into vegetation, soil conditions, and overall ecosystem health.

Indicators of biodiversity and soil health—such as plant variety, soil composition, and microbial activity—can now be monitored in detail.

AI helps analyze these factors, providing a clearer picture of the farm's ecological status.

With real-time feedback, farmers can make quick, informed decisions to improve their operations.

Adaptive management allows them to tweak their practices as needed, supporting both productivity and biodiversity.

AI is also being used to optimize fertilizer and pesticide use.

Machine learning helps analyze crop and soil data, making it easier to apply inputs precisely where and when they're needed.

This reduces waste and enhances crop performance.

Predictive models allow farmers to forecast pest outbreaks and nutrient deficiencies.

By acting proactively, they can avoid overuse of chemicals and improve input efficiency, promoting sustainable farming practices.

This has positive effects on both the environment and farmers' bottom lines.

Less chemical input means healthier ecosystems.

And with lower input costs and better yields, farmers can boost their profitability while farming sustainably.

Let's move on to precision carbon farming.

Advanced sensors are now able to measure soil organic carbon with high accuracy.

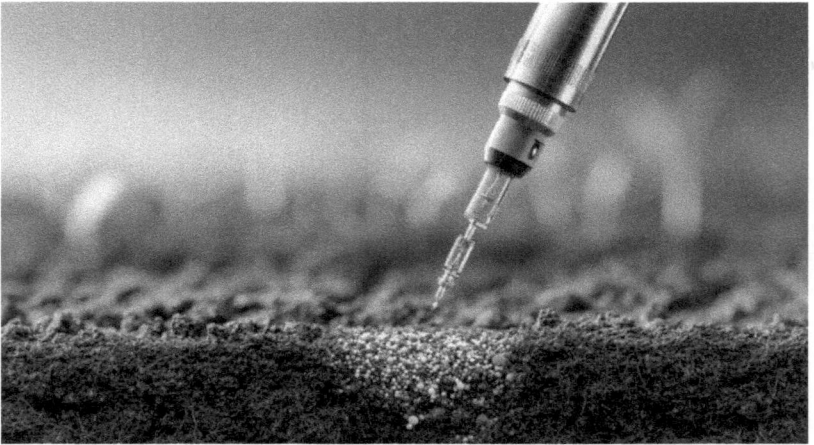

AI processes this data, helping track carbon levels and dynamics in the soil.

Chapter 8: Sustainable and Regenerative Agriculture

With AI insights, farmers can manage soil carbon more effectively.

By recommending specific soil amendments, encouraging cover cropping, and minimizing tillage, AI supports strategies that enhance carbon sequestration and improve soil health.

AI also plays a role in verifying carbon sequestration.

With accurate measurement, farmers can access carbon credit markets, creating new income opportunities.

This not only incentivizes sustainable practices but also contributes to climate change mitigation.

To conclude, AI is proving to be a powerful ally in sustainable and regenerative agriculture.

By supporting smarter farming practices, enhancing environmental stewardship, and offering economic benefits, AI is paving the way for a more sustainable future in agriculture.

Chapter 9: AI in Food Production and Processing

Let's start with a broad overview of AI in food production and its implications for the industry. AI leverages advanced data analytics to help manufacturers make better decisions based on real-time data. Machine learning models can predict trends, improve food quality, and enhance safety. Plus, automation powered by AI streamlines production processes, making operations more efficient than ever before.

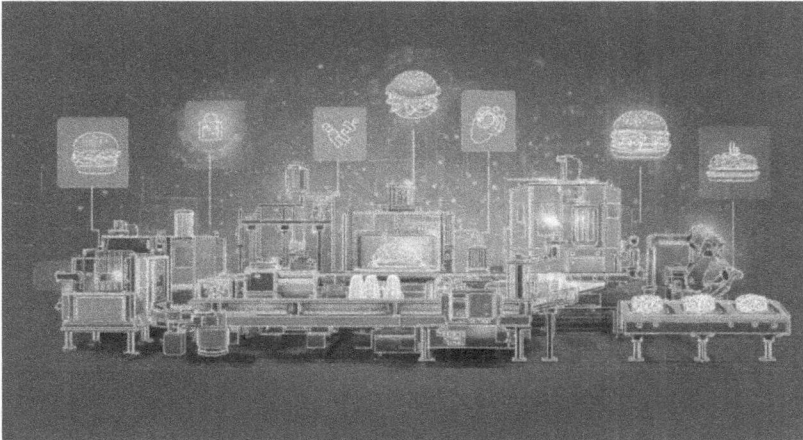

There are significant benefits to adopting AI in the food industry. It increases efficiency by automating repetitive tasks and improving process flows. AI also boosts quality control by detecting defects early and ensuring products meet consistent standards. On top of that, AI helps reduce operational costs by optimizing resources. However, adoption isn't without its challenges—companies often face issues like data integration, high upfront investment, and the need for workforce adaptation.

Emerging trends show AI being used for real-time monitoring of production lines, ensuring safety and consistent quality. Predictive analytics also plays a big role, helping businesses anticipate demand and optimize supply chains. And as AI continues to drive automation, food production becomes more efficient, reducing the need for manual labor while increasing output.

AI-powered sensors can monitor equipment conditions continuously, providing real-time data for more accurate tracking. These systems can detect anomalies early on, preventing unexpected failures. And with insights from AI analytics, companies can schedule maintenance at the right times, minimizing downtime and keeping production running smoothly.

Predictive failure detection is a key advantage of AI in food plants. By forecasting potential equipment failures, companies can act proactively to avoid costly disruptions. This predictive approach also helps maintain a steady workflow, ensuring that production schedules remain on track and customer deliveries are not affected.

Predictive analytics doesn't just reduce downtime—it also cuts repair costs by catching issues before they become serious. AI monitoring systems contribute to safer workplaces by detecting equipment problems early and preventing dangerous malfunctions. This proactive approach supports both operational efficiency and worker safety.

AI plays a powerful role in developing new recipes and formulations. By analyzing data, AI provides valuable insights that guide recipe adjustments and innovations, helping companies create better, more appealing products.

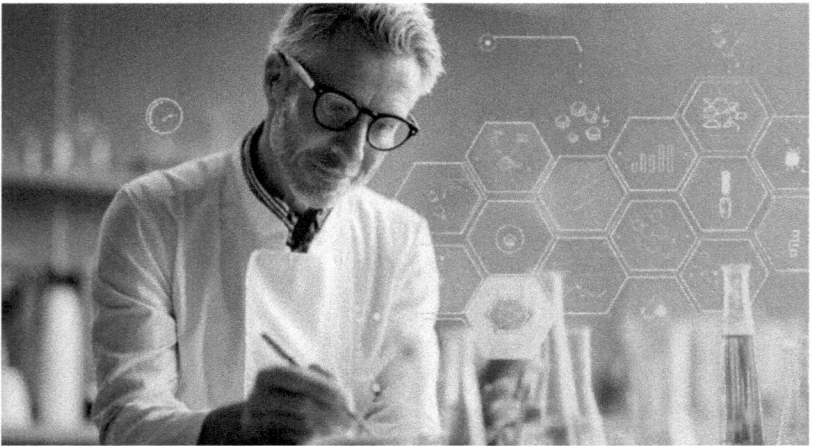

AI can assess nutritional content and suggest ingredient substitutions to meet specific dietary goals. Whether it's for health reasons, allergies, or cost considerations, AI helps fine-tune recipes while maintaining or even improving the end product's quality.

Personalization is another area where AI shines. Machine learning analyzes consumer data to create food products tailored to individual preferences and needs. This means addressing dietary restrictions, allergies, or personal taste, all of which enhance customer satisfaction and loyalty.

Robotics is revolutionizing food packaging and handling by automating repetitive, labor-intensive tasks. Robots sort products with precision, handle packing operations efficiently, and manage palletizing with speed and accuracy. These innovations increase productivity and reduce reliance on manual labor.

Robotic systems also enhance hygiene in packaging environments by minimizing human contact, which reduces contamination risks.

Automation ensures that every product is packaged consistently, maintaining high standards of quality with every batch.

Let's look at some examples of robotics in action. Robots can significantly boost throughput, making food plants more productive. They also help reduce human errors, leading to better consistency and quality control. Finally, by taking over hazardous tasks, robots make workplaces safer for employees.

AI-powered tools are now used for sensory evaluation, going beyond human capabilities. Advanced sensors collect detailed data, while AI algorithms analyze this data objectively—eliminating human bias and providing more consistent results. These tools complement traditional taste tests, giving a fuller picture of a product's sensory qualities.

Machine learning models can predict consumer taste preferences by analyzing sensory data. By correlating feedback with specific attributes like flavor or texture, AI helps guide product development to better meet customer expectations.

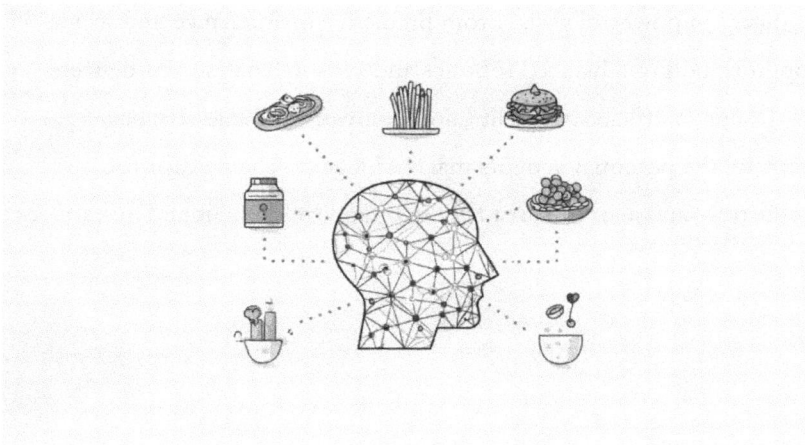

Advanced sensory analysis using AI enables companies to refine product quality in targeted ways. Whether it's adjusting a recipe for better flavor or tweaking texture for enhanced appeal, sensory AI provides actionable insights that drive product improvement and increase consumer satisfaction.

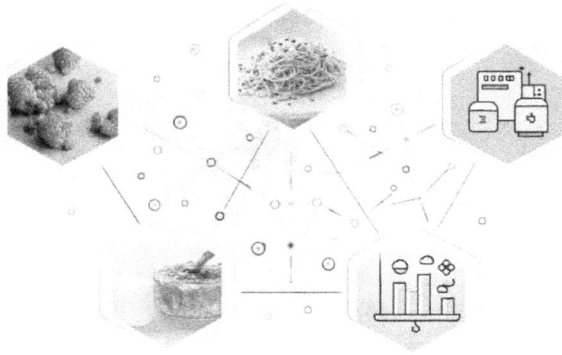

To wrap up, AI is reshaping the food production and processing industry in powerful ways. From predictive maintenance and recipe optimization to advanced robotics and sensory analysis, AI delivers real value in efficiency, quality, and innovation. While challenges remain, the potential benefits make AI a game-changer for the industry—and a critical area for ongoing development and investment.

Chapter 10: Food Security and Policy Forecasting

Food risk modelling frameworks integrate data from diverse sources—such as climate data, crop yields, market prices, socio-economic factors, and conflict data. By combining these, models improve the accuracy of food insecurity predictions. They use advanced analytics and algorithms to assess and forecast food risk scenarios, helping identify potential crises before they happen. These insights guide policymakers and humanitarian agencies in making timely, proactive decisions.

Effective food risk assessment depends on using a wide range of data. By integrating climate patterns, market conditions, socio-economic data, and conflict reports, models can capture the complex factors that drive food insecurity. The key takeaway here is that the more comprehensive and integrated the data, the more accurate and relevant the risk predictions will be.

These models also have practical applications in identifying vulnerable regions. By mapping food insecurity hotspots, we can pinpoint areas needing urgent support. This enables targeted interventions, ensuring that aid reaches those who need it most. Moreover, governments can use these insights to optimize resource allocation and minimize risk across different regions.

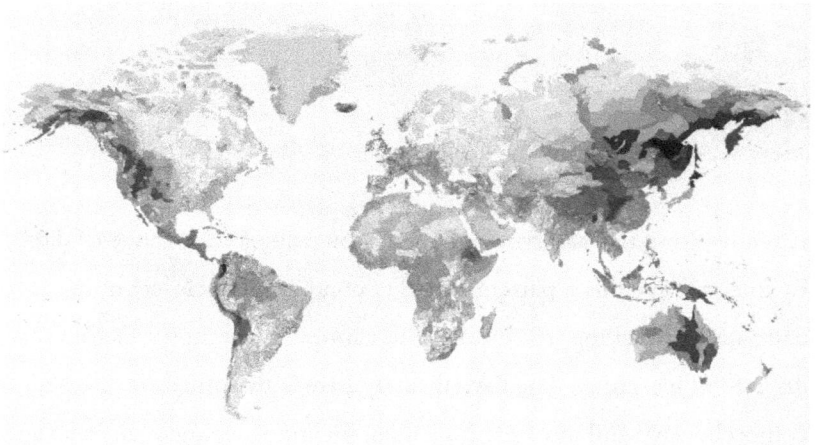

AI is transforming agricultural policy simulation. With machine learning and agent-based modeling, we can now run dynamic simulations that analyze the potential impacts of policy decisions. AI allows for scenario testing, so policymakers can evaluate different strategies and foresee possible outcomes under varying conditions before taking action.

When building these simulation models, careful design is essential. Models need to incorporate both agricultural and socio-economic variables that impact food security. Calibration with real-world data is critical to ensure models remain accurate and reliable. It's also vital that the models reflect the complex interactions within food systems that affect food security.

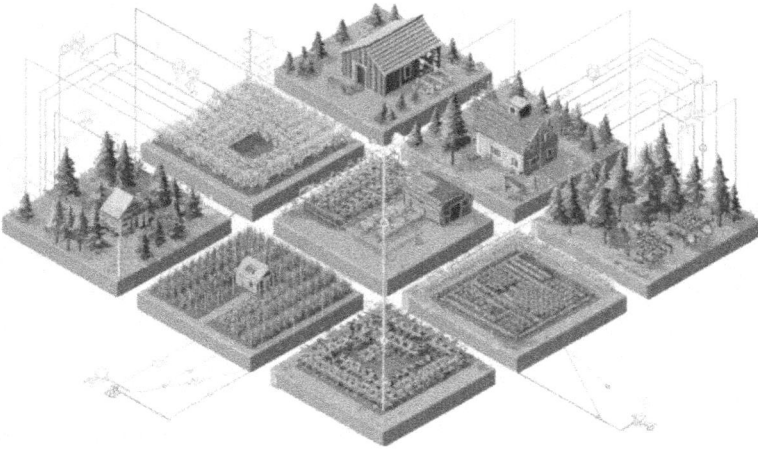

Simulations give policymakers a powerful tool to predict the outcomes of their decisions. By forecasting possible impacts, they can refine strategies to enhance food availability, accessibility, and stability.

Essentially, simulations help minimize risk and improve policy effectiveness before real-world implementation.

Advanced techniques like machine learning are improving large-scale crop yield prediction. By analyzing complex data patterns, these algorithms enhance forecast accuracy. Remote sensing, using satellite imagery, provides large-area monitoring of crop health, while statistical modeling brings together various data sources to guide both farmers and policymakers in making informed decisions.

Satellite imagery offers broad insights into crop health and environmental conditions over time. Ground sensors complement this by providing real-time, detailed local data. When we integrate satellite and sensor data, the result is improved prediction accuracy—allowing better assessment of both crop yields and environmental factors.

Accurate yield prediction plays a crucial role in effective aid allocation. By knowing where food shortages are likely to occur, we can target aid efforts more precisely. Data-driven allocation helps ensure that resources reach those who need them most, reducing waste and maximizing the impact of food aid programs.

Monitoring geopolitical events is key to understanding disruptions in food supply. By tracking conflicts, we can anticipate impacts on food production and supply chains. Analyzing trade disruptions helps identify potential delays or increased costs that affect food availability. Additionally, keeping an eye on policy changes allows us to foresee regulatory shifts that may impact food imports and exports.

Real-time monitoring of climate stressors is critical for timely response. Continuous environmental data collection helps detect events like droughts, floods, or extreme weather conditions that can threaten agriculture. Identifying these events early supports rapid response efforts to protect food production and sustain ecosystems.

By integrating diverse risk data, we can develop responsive strategies for food systems at risk. Policy tools can be used to craft adaptive strategies that address emerging threats. These strategies help build resilience in food systems, enabling them to withstand shocks and stresses. Importantly, they also focus on protecting vulnerable populations, ensuring equitable access to food during times of crisis.

To wrap up, we've seen how advanced modelling, AI simulations, yield prediction, and real-time monitoring are revolutionizing food security and policymaking. These approaches support better decision-making, enhance resilience in food systems, and help protect vulnerable communities. By leveraging data-driven insights, we can proactively address food security challenges in an increasingly complex world.

Chapter 11: Market Intelligence and Agribusiness Innovation

Data-driven decision making is critical in modern agriculture. By leveraging data, farmers can enhance productivity by optimizing how they use resources and manage crops. They can also better anticipate risks like weather changes or pest outbreaks, protecting their yields. Plus, data-driven farming promotes sustainability by reducing waste and environmental impact through precision techniques.

Technological advancements are reshaping agribusiness. Precision farming uses data and automation to improve yields and sustainability. IoT sensors collect real-time information on soil, weather, and crop health, supporting smarter decisions. Drones provide aerial insights and enable targeted input application, while blockchain enhances transparency and traceability across the agricultural supply chain.

Market intelligence plays a vital role in shaping strategies for aggrotech businesses. By understanding consumer demand, companies can create products that truly meet market needs. Monitoring competitor activity helps maintain a competitive edge, and identifying emerging trends allows businesses to innovate and stay ahead in a dynamic industry.

Machine learning is revolutionizing price forecasting in agriculture. Models like regression, decision trees, random forests, and neural networks analyze historical data to predict price movements. These algorithms help businesses understand price patterns, leading to better decision-making in the market.

Accurate price forecasting requires high-quality data and thorough preprocessing. Clean, relevant data allows machine learning models to generate meaningful predictions, making them powerful tools for agricultural markets.

Real-world case studies show how machine learning price forecasts can help farmers decide the best times to sell their crops, maximizing profits and minimizing risks. These examples also highlight the economic benefits for agribusinesses that adopt such technologies, leading to smarter market strategies and better financial outcomes.

AI is a powerful tool for understanding customer needs in aggrotech. By analyzing customer feedback, AI can highlight unmet needs and guide product development. Studying purchasing patterns helps predict future preferences, while mining social media offers real-time insights into consumer trends.

AI-driven insights optimize product development. By predicting market response, companies can better tailor their offerings. Simulating product performance before launch reduces risk, and leveraging AI shortens development cycles—helping companies get products to market faster and more effectively.

Success stories in aggrotech show that AI helps businesses enhance product-market fit, leading to higher adoption rates among farmers and greater customer satisfaction. By aligning solutions with farmers' real needs, AI-driven innovation supports stronger market engagement.

Large language models, or LLMs, are emerging tools for evaluating aggrotech startups. These models can analyze vast amounts of market data, making them valuable for understanding startup value and potential in the evolving agribusiness landscape.

LLMs help detect and interpret market trends. By analyzing consumer sentiment, tracking innovations, and identifying growth opportunities, these models enable businesses to anticipate market shifts and adapt their strategies proactively.

Advanced analytics powered by LLMs enhance investor decision-making. These tools provide deeper insights into market dynamics, offer predictive capabilities for trend forecasting, and improve risk assessments—making them crucial for informed investment choices in agribusiness.

Fintech is addressing agriculture's unique financial challenges. Farmers gain access to financing solutions tailored to their seasonal income cycles. Digital payment platforms simplify transactions, while credit scoring models consider agriculture-specific factors like weather and harvest timing, making financial services more accessible.

Technology is also transforming agricultural insurance. Weather-indexed policies and parametric insurance allow quicker, more reliable payouts based on predefined conditions, giving farmers greater financial security and reducing claim processing times.

Looking ahead, financial services in aggrotech will increasingly integrate AI, blockchain, and IoT. AI will enhance risk assessment and decision-making. Blockchain will improve transparency and trust in financial transactions. And IoT devices will gather critical real-time data, boosting efficiency and service delivery in agricultural finance.

To conclude, advanced technologies like AI, machine learning, LLMs, fintech innovations, and IoT are driving a major shift in the aggrotech sector. By harnessing these tools, agribusinesses can make smarter decisions, innovate effectively, and build more resilient, sustainable models for the future.